Google Glass: What Is It and How Can It Change Our Lives

Legal Information

Copyright 2013 Future Apps
Published in America

All Copyrights and trademarks found in this book are property of their respective owners.

Table Of Contents

Page 4 – Introduction

Page 11 - The Story of Glass

Page 13 - The Google Glass Explorers

Page 35 - Is Glass Socially Accepted?

Page 38 - The Glassware Apps

Page 59 - Augmented Reality

Page 73 - How Google Glass Handles Your Privacy and The Privacy Of Those Around You

Page 77 - Google Glass and The Connected World

Page 82 - Google Glass Will Change How You Share Forever

Page 87 - What Goes Into Glass?

Page 90 - Google Glass is Only Just Beginning

Introduction

What is Google Glass anyway? What makes it so special?

Well the answer is actually quite simple. Google Glass is a pair of glasses with a small transparent screen attached. On this screen Google Glass can display a variety of useful information, such as step-by-step map directions, or the local weather.

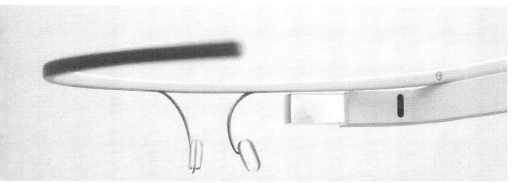
The small piece of glass is where all the information is displayed.

And all of this is simply right there, available at all times, right on the top of your vision.

But it goes much deeper than that. Glass also holds a camera, microphone, and a small speaker. This opens up a whole new world of amazing things which this device is capable of!

You can easily and quickly capture video just by saying "Okay Glass, record video". Or message your friends in much the same way.

When it comes down to it, Google Glass really is a glimpse into the future, the future of not only technology but also the entire connected world.

The Camera

Glass takes a photo of what you can see.

"Okay Glass, Take a Photo", and that moment is captured forever. You don't need to worry about fumbling with your phone or digital camera. You don't have to wait for it to power up. Google Glass is ready at a moments notice. Just align the picture using your own eyes.

What ever you see, Glass sees. This makes taking photos or videos, dead simple. You never miss something that is out of frame because you had to hold your phone out.

Google Glass takes the physical barrier out of technology. It brings the world one more step closer while still keeping up all the benefits that new connected and social technology has to offer.

Let your friends see what you're doing while you see them.

Want to share what you're doing with your friends? Google Glass will instantly stream what ever you see to anyone you want. You could be at a concert while your friends and family at home catch their favorite song all because you streamed it live to them.

It's just one more way Google Glass connects everything – and everyone together.

Maps

Looking at a paper map is slowly becoming a thing of the past, with smart phones and GPSs making them irrelevant.

Google Glass is rapidly making paper maps a thing of the past.

Never again get lost without directions.

You can get step-by-step directions that match up with what you are seeing.

It doesn't matter if you are driving, using public transport, walking, hiking, or biking. Google glass shows you how to get to anywhere you want.

Glass Is Your Personal Assistant

You can ask Glass just about anything and it will answer! Find out what you want to know simply with your voice.

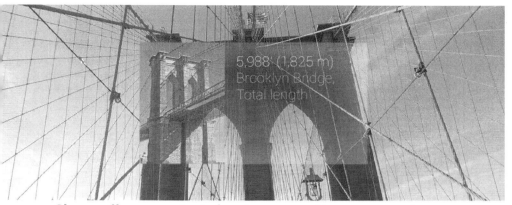

Glass will answer just about any question you may have.

No matter what you want know Google Glass will have an answer. You can ask anything from questions about historical buildings to how your favorite sports team is doing to questions about your calendar!

Never again will you have to go without knowing something. Google Glass makes it not just quick but easy to access the rich world of information all around us.

It quickly becomes apparent how Google Glass could fundamentally change our lives and our world.

It's not exactly about what Glass does. It's really more about how it does it. Glass brings information so quickly and seamlessly to our lives

Glass with update you flight delays or boarding gate changes.

Without even asking Glass presents extremely useful and relevant information straight to you. If you need to know what boarding gate to go to at an airport, Google Glass is already showing you what you need to know.

The same is true for a massive variety of other places and other pieces of information. The limitless applications for glass could change the fabric of our living forever.

Glass can update you on a traffic accident and change your route. It can let you know how much time you have until your movie begins or where the closet Italian restaurant is.

You can use glass to capture the world around you and share it with your friends and family.

Glass can become your personal assistant with highly important information or just make your life easier by removing the physical part of technology. Freeing your hands and your eyes.

The best part is, we've only scratched the surface of what Glass is and what it can do!

The Story Of Glass

Google Glass was born years ago in the secret off-campus Google R&D center known only as Google X.

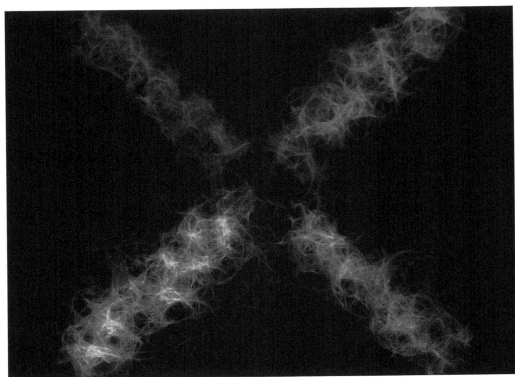
Google X is known for it's famous projects, like the Google self-driving car.

For the longest time no one outside of Google even knew these prototype glasses existed. Until one day in late 2011 the story finally broke. Technology news outlets everywhere picked up the story that Google was working on some type of "wearable computing device".

Technology enthusiasts everywhere went wild. All over the Internet people were posting how these glasses could be used in their every day life.

The theories were all over the place, some people were insisting that the future had finally arrived. Technology was finally going to reach the level only seen in Sci-Fi movies.

Several months later Google finally came out with what they were working on, they called it: Project Glass.

Better yet, they released an official video of what Google Glass could do. And it blew everyone's expectations out of the water.

It was a video of an average person going through their daily lives while wearing Google Glass.

He starts his day off by checking his calendar for his plans for the day. Just up looking up at Google Glass display he instantly sees that he is meeting up with "Jess at 6:30 tonight".
After making his morning coffee he approaches his apartment window and looks out over the New York skyline. Google Glass is telling him that it's going to be 58 degrees and cloudy today with a 10% chance of rain.

While having breakfast his friend sends him a text that instantly pops up on Google Glass: "Wanna meet up today?" Just by using his voice he texts back asking him to meet up by the local bookstore

As he approaches the subway stairs to make his way over to the bookstore, a notification pops up on Google Glass: Subway service suspended.

Live notifications about the world around you.

Google Glass doesn't skip a beat and instantly shows him the walking route directions.

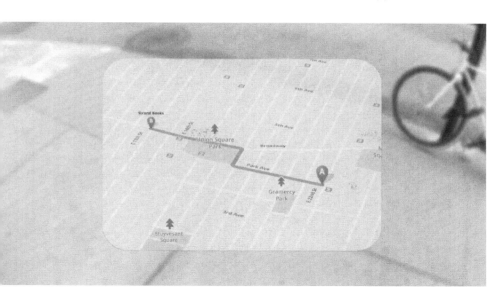
Turn by turn directions whenever you may need it.

On his way to the bookstore he sees a poster for a local band playing tonight and asks Google Glass to remind him to get tickets.

Once he gets to the bookstore he asks Glass: "where's the music section". Google Glass not only responds with a map of the store and where the music section is, but also with the fastest walking route.

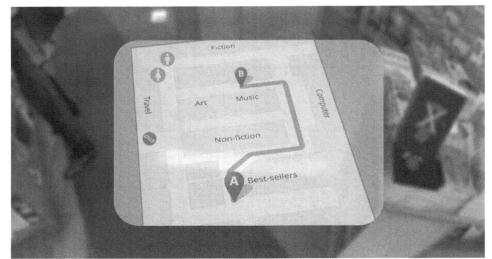

Google Glass can show you indoor maps for malls and stores filled with all the useful information you would ever need.

Immediately after he finds what he was looking for he asks Glass: "Is Paul here yet". Google Glass then asks his friend Paul to share his location and Google Glass displays how close he is.

Friends and family can share their location and see where you are.

He meets up with his friend and they go eat at the location food truck. He checks in his location with Glass and chats with his friend.

Later on after saying goodbye to his friend he sees some graffiti art he likes and asks glass to take a photo.

Glass reminds him that he's meeting up with Jess soon. He then turns off his music by telling Glass: "Music Stop".

At this point he's on the rooftop and accepts a video chat with Jess. He can see her on Glasses display and she can see the sunset he is looking at.

Video chats can become much more than simply talking to someone else.

And that is the end of the video. It showed off practically everything Glass could do and how it could add some much to anyone's everyday life.

From the useful notifications about the subway being closed for maintenance to the indoor map of the bookstore Google Glass is always there when you need it and without getting in the way.

But that was only the start. At Google's yearly I/O conference they had skydivers wearing Glass stream everything they were seeing down to the conference live.

Glass can even stream live from a skydive.

Everyone in the audience experienced the jump for themselves just by watching it. After the jump they also showed Google Glass on mountain bike riders and mountain climbers.

Video footage from Google Glass

It goes to show how connected Google Glass makes everything. No matter what you're doing Glass is always with you, ready to record at a seconds notice.

It may just be that because Glass is so much more convenient and easy to access that we can actually start capturing the coolest and best moments of our lives unlike ever before.

Captured on Glass

One day we might be seeing this very often.

One other amazing piece of information came out of Google's I/O conference: The Glass Explorer Edition Program.

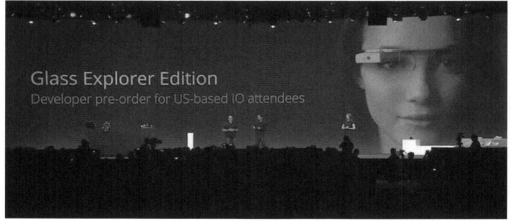

The Google Glass Explorer Edition revealed at Google I/O

The Glass Explorer Edition Program is Google's open public beta test for Google Glass. Virtually anyone could get selected to get their hand on their own pair of Glass.

All they needed to do was tell Google why they could showcase Google Glasses abilities by exploring the world. Thousands of people applied yet only a few got accepted.

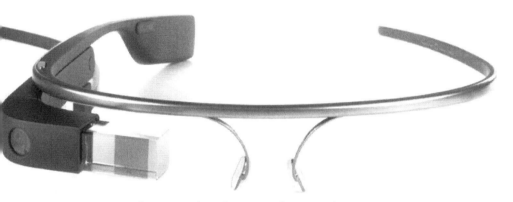

The Google Glass Explorer Edition

These early beta editions of Google Glass came with a hefty price tag. If you were accepted you would need to cough up $1500.

This crazy price was mostly to get only serious candidates, people who would be willing to pay the price to get the very forefront of cutting edge technology.

These early adopters would have the chance to "shape the future of Glass".

Nine months later the first explorer edition pairs of Glass went out. People all around the world started uses them and testing them, seeing what they could do.

The early adopters of Google Glass would be able to shape its future

The amazing thing was that everything was living up to the original hype. Google Glass could do what it promised and that could change everything. Read on to find out what they had to say.

Finally Google officially commented that they are aiming for a public release sometime in 2014.

Google also announced that it would be at a much, much reduced price from the explorer edition.

The Google Glass Explorers

How they may decide the fate of Google Glass

The Google Glass Explorers received their own pair of Google Glass in April of 2013.

Once they completed their sign up and paired the glasses with their Google account Glass was ready to go.

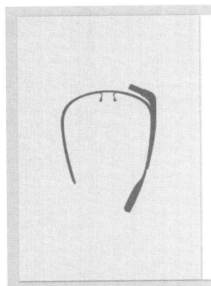

The sign up page that the explorers saw when activating Glass

They had to pair them with their android smart phones for certain features to function but other than that requirement everything worked as expected.

Initial Reactions

The initial reactions to Google Glass were all positive, "Everyone who puts Google Glass on for the first time immediately lights up with a smile".

A Glass Explorer wearing his own pair of Google Glass

Overall people were expecting a lot out of this device and they were hoping that glass could deliver.

As people got more and more use out of their Google Glass is became clearer what this device could do well and what it couldn't.

The Positives

The Screen

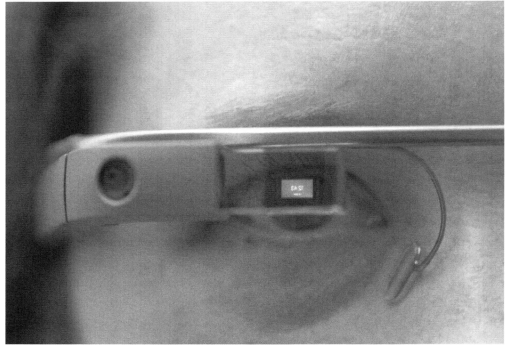
The high quality display on Google Glass

The display on Google Glass is crystal clear; it looks like a screen that just floats in front of your eye.

It may seem tiny in the photo above but when you wear Glass the screen is so much closer to your eye that it seems a few inches in size.

From all the reactions from the Google Glass explorers the screen always blows them away.

Camera

A photo taken by Google Glass

The camera isn't the highest quality and it doesn't have build in flash, but it makes up for that in it's utility.

Google Glass takes photos the way photos were meant to be taken. You "frame" the picture by lining up what you see and use your voice to take the photo.

It takes a bit of practice but once you've got the hang of it you won't ever need to whip out your cell phone camera again.

A Google Now card with the game's final score

Google Now has long been a feature on Android smart phones. It gives you relevant information without you asking for it.

It's built to be smart and it really is. Google Now will remember when you leave for work or school and give you the traffic information before you ask for it.

And that's just the tip of the ice burg Google Glass and Google Now go hand in hand.

Maps

Turn-by-turn directions on Google Glass

Maps are a daily part of our lives; smart phones have made them more and more useful and now Glass takes that even further.

By overlaying the maps over part of your vision, Glass makes it easier than ever to keep track of where you need to go. The maps will even turn around as you move your head around, keeping track of your orientation.

This is yet another case where you won't need to take out your smartphone and instead use Google Glass.

Notifications

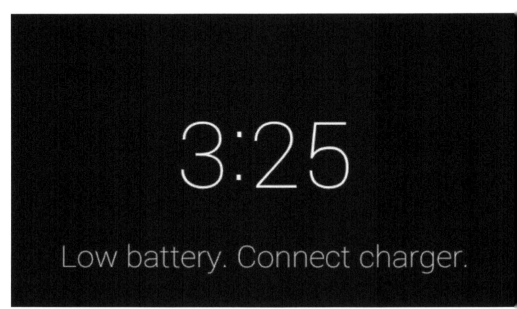

The main screen not only displays the time but also any notifications

The notifications on Glass are instant and extremely useful. Instead of feeling your smart phone vibrate in your pocket and having to guess at what that could you, all you need to do is look up at Google Glass.

Google Glass displays notifications for a wide variety of things, from incoming calls, to texts, and even emails.

It's another case where Glass takes away the barrier between technology and you. Nothing could be simpler than already having the information you want in front of you.

Apps

The New York Times Glass app in action

Google has finally granted all 3rd party developers access to Google Glass, and they have designed some great apps.

Right now, even just for the beta edition of Glass there are over a 100 specially made apps, everything from Evernote to the New York Times.

Even more apps will become available before Google Glass becomes available to the general public. Read on for much, much more on Glass' apps!

The Negatives

Typing

An Android smartphone keyboard

Just like with all voice recognition software: it isn't perfect. And with Google Glass you need perfect fro two reasons.

First of all because it's incredibly difficult to fix what you wrote, you'll most likely just need to repeat everything over again.

And second, because for a lot of apps, like messaging, Glass will send what ever it thinks you said (even if it's wrong) right away, without asking for the final okay.

That's why you're better off using your phone; at least until voice recognition gets much better.

Limited apps

You can't read very much on the smaller screen

The smaller screen of Google Glass has its advantages but it also has its disadvantages. The biggest one being that not much text will fit on the screen all at once.

This really limits certain apps such as Gmail or The New York Times. It's great for shorter messages but anything longer than a few sentences makes reading a bit tedious.

Glass won't be replacing your E-reader any time soon.

How Explorers Can Help Change Glass

The Explorers have a unique opportunity to have change the face of glass into something that they just love using.

That's because Google is listening to their feedback every step of the way. They take into consideration everything they like, don't like, and want changed and make their decisions based on that.

This will lead to Google Glass ever evolving and changing into the best product it can be... And all before the public release.

The Google Explorers are also going to shape the public opinion of Google Glass and so it's not unlikely that they can in fact decide if Google Glass will be a success in the end.

Glass is built directly into all social media, with simple photo and video posting, status updates and more

If they show the world just how amazing Glass is, then they might shift things in favor of Glass being a massive success.

A lot rests in the hands of the Explorers.

Is Glass Socially accepted?

One thing that almost every single Google Glass Explorer noticed is how popular they became while wearing Google Glass.

Everyone was looking their way, checking out that stylist futuristic piece of technology they were wearing.

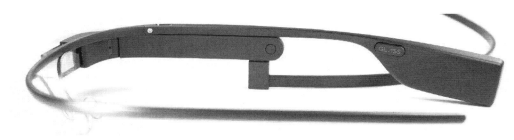
Google Glass from the side

By far the most common thing they were asked, over and over again, was: "Is that Google Glass?" and "what is that?"

Over and over they had to answer the same questions. They become were versed with their answers.

With this many people looking at the Explorers and asking them questions it begs the question: How long will it take for Google Glass to become socially accepted?

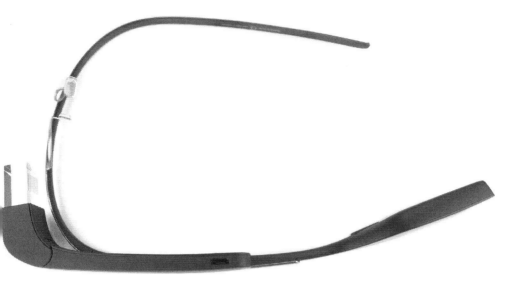
Google Glass from the bottom

Even after years of "voice assistants" like the iPhone's Siri have become popular most people still don't like talking to their phones in public areas.

Google Glass magnifies this problem ten fold, as the main way to interact with Glass is to talk to it.

So how long will it be until Google Glass is so socially accepted that no one even blinks an eye when someone is walking around wearing Glass and seemingly talking to his or her glasses?
It might be here sooner than you think. It didn't take very long for cell phones to become socially

accepted and that was once viewed as a very different thing.

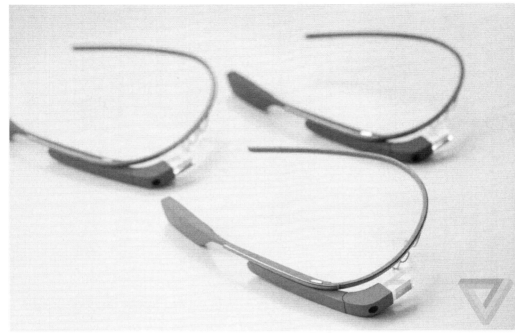

The different color of Glass

There was a point where everyone with a cell phone got the looks and stares of strangers.

The day where Glass just blends in is getting ever closer. We are slowly moving into the future, a future where Glass is just another piece of tech.

Evernote

The Evernote app on Google Glass is a simplified version of the mobile app. You can use glass to post photos straight to your notes.

You can also dictate and post them to your notebooks.

There is support for shared notebooks as well. However there isn't much in the way of creating new notebooks at the moment.

New York Times

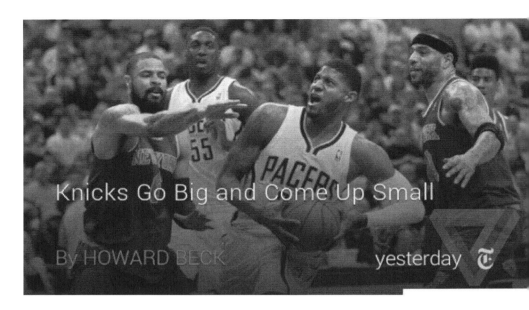

The New York Times app is a much simpler version of their website, MUCH simpler.

The app sends you a new "card" every hour or so with all the latest headlines and news pieces that have been posted.

You can scroll through the stories and if you want Google Glass can even read them to you.

There's no functionality to search or browse through the categories. However the app is great for staying on top of current news.

CNN

The CNN app is very similar to the New York Times app (as they are both news apps) however the CNN app is superior in a variety of ways.

More customizable options is really what sets the two apps apart.

With the CNN app you can decided how often you receive a notification and you can choose which topics you'd like to hear from.

CNN will also deliver video news to your Google Glass.

Tumblr

Tumblr lets you upload photos to your Tumblr. You can upload as many as you want.

You can also access your own Tumblr timeline. Unfortunately your timeline will look extremely flooded if you follow a lot of Tumblr's on Glasses small screen.

Google Glass can only view one photo at a time so getting through your timeline is a little tedious. It does limit the usefulness of the app.

But it's a small thing that won't matter to everyone.

Facebook

The Facebook app is limited to sharing photos to your timeline. You can share you photos with everyone or just yourself.

You can also add descriptions to anything you've uploaded.

However you can't post comments, or like things.

You also can't upload video; right now all you can upload are photos.

Twitter

The Twitter app lets you tweet photos. For some reason that's it's only ability.

There's no way to just tweet text, but you can respond to other tweeters with text.

So far the twitter app is great if you just care about posting photos.

If you are a text only tweeter, the Glass app doesn't have what you need.

Glassagram

This app will let you alter your photos with a variety of "instagram" style filters.

It's very simple with only a few different styles to pick from. But it is a great start if you like changing your photos up to add that personal style.

After you alter your photos you can upload them to any of your favorite social media websites using their apps.

Glassnost

Glassnost is a Glass only style social media website. Where people with Glass can share their photos with other glass users.

For now you can only upload photos taken with glass, no videos.

You can also like other people's photos and comment on them.

Glass to Do

This app lets you create a variety of to-do lists that are saved to your timeline as cards.

It's useful and works as described. You can check off your to-dos as you go along.

So if you are someone that loves to create a to-do list this app is for you.

The only downside is that the app is limited and only lets you create very simple to-do lists, nothing else.

Elle

The Elle app will give you an update on whatever news you want to receive from the fashion magazine.

You can change what type of news you'd like to receive and how often in the apps preferences.

This app works a lot like the CNN app and can even read out the different pieces of fashion news to you.

Full Screen Beam

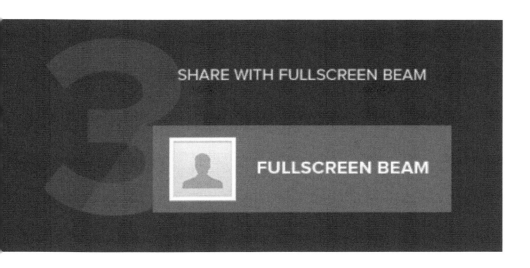

Full Screen Beam is the only app that lets you upload videos straight to YouTube.

Any video that you have recorded can be uploaded. You can add a title and a description.

Once you've uploaded your video you can then tweet the link out to your followers.

This app is great for people who want to upload their videos straight to YouTube and for people who just want the ability to upload video with Google glass.

facebook

This app beats Facebook at it's own game. With this app you can basically do everything the original Facebook app couldn't.

You can upload photos AND update your status.

Want to reply to a comment? With this app it's simple and easy.

You can even like the different photos or status updates you come across.

In almost every way this app is better than the official one.

Thirst

The thirst app is yet anther news app but with a twist. This app takes all the trending content on social media websites and delivers them to you.

You can pick and choose what type of news you'd like to receive.

The quality presentation is what sets it apart from the other news apps. The app can read an entire story to you.

All in all this app has all the features you would look for in a news app.

Google Search

You can Google things on your Google Glass just by asking it with your voice.

It works best if you ask it mostly questions as it has great answers to most questions about almost anything you can imagine.

However the search isn't perfect, if your question is too complicated you will just see a simplified version of the Google search results. Which does not look good on Glass.

Worst of all is you can't click on any websites for more information.

Take a Picture

This app is the lifeblood of Glass, taking a picture is quick and easy.

You just say: "Okay glass, take a picture" and it does.

Sharing the photo is a matter of choosing your favorite social media app and sharing it though that.

The standard is Google Plus, however that's not nearly as popular as Facebook or Twitter.

Record a Video

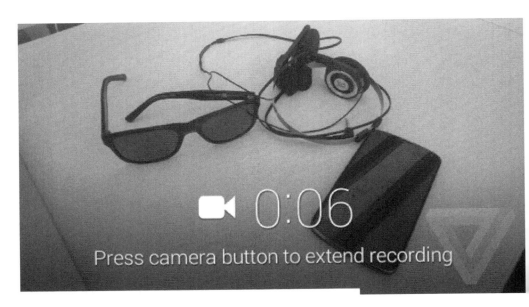

You can record short videos using Google Glass. The longest video now supported is 10 full seconds.

But that's bound to change eventually. The video app is great for capturing any moments that you may want to share on YouTube or to other places.

Right now the app is very simple with no editing tools or built in sharing features. But you can always use other apps for that.

Google Maps

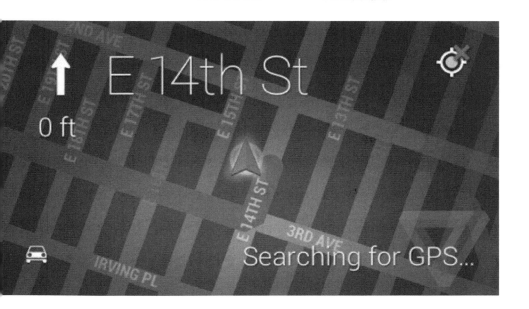

The Google Maps app is extremely useful. It's still a simpler version of what you would find on your phone or computer but it works so well it doesn't even matter.

The turn-by-turn directions make finding your way around any city a breeze. You don't need to constantly look at your phone or decipher maps; instead you only need to look up at Glass.

There's support for walking, cycling, and driving. There's no support for public transportation right now but that's sure to arrive in a future update.

Messaging

> :Mark Gurman
>
> Watched your video. Amazing work man
>
> yesterday

With the messaging app just saying, "send a message to…" will bring up a list of contacts for you to pick from.

Writing the message is as easy as talking to Glass, just don't make any mistakes, as it's almost impossible to change after the fact.

Overall the messaging app works well enough, you can only use Gmail right now but in time that should expand to include all major email providers.

Phone Calls

You will need to have your Android smart phone paired with Glass to actually make phone calls but once you have that set up making calls is a breeze.

You can call any of your phone contacts just by saying "call..." and their name.

The only issue with Glass is that the bone conducting audio might be too quite for most people.

Google Hangouts

> ok glass, hang out with...
> Chad Mumm
> Jordan Oplinger
> Joshua Topolsky
> Billy Disney

Google hangouts is Google's brand new messaging service. You can message your friends and family separately in group messages.

You can also use Google handouts to video message anyone in your contacts.

This also lets you share what you see using Glass' camera.

Overall the messaging works well, with slight problems with speech recognition.

Google Now

Google Now is almost the same as the Android version. So like the Android version Google Now works pretty well but not all the time.

It's great for location-based things such as traffic estimates and airplane information at the airport.

But sometimes it doesn't do much at all. The hope is as Google Now matures, more and more features become available and more useful on Google Glass.

Augmented Reality
What it is and why it might be the most important thing for Glass

Beyond all the apps and features the one thing that Glass has that could just tip everything in its favor is augmented reality.

Just what is augmented reality? Well it's adding more information, usually using some kind of screen to the world.

Augmented reality technology usually uses a camera to figure out what is around it and it uses that info to show you relevant info.

With Glass this becomes so much simpler, as the camera is attached to Glass and the screen is right on top of your vision.

It could be as simple as looking at a bus stop and having glass show you how much time until the next bus shows up, and which one.

Or it could be used in museums or art galleries where just looking at a piece of art shows you more information.

There are so many possibilities, so much can be done with augmented reality and that is what makes Google Glass so exciting.

The best part is that this type of innovation is already happening, there are many companies moving toward a fully connected future.

A great example is the airline JetBlue, when they submitted their entry into the Google Glass Explorer Program they used these mock ups showing what they would try to achieve.

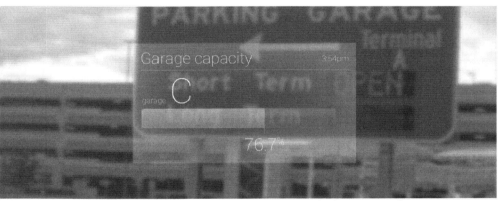
Google Glass displaying how much space is left in a parking garage

The first picture displays how much space is left in the airport parking garage, it even displays which parking garage it's talking about.

This type of technology would make it much easier to find available parking, not just at airports either but anywhere with parking garages.

It would be a major help at massive sporting events or concerts. Augmented reality could even help you find your parking spot.

Google Glass showing the location of available outlets around an airport waiting area

This second picture shows you how Google Glass could augment reality and help you find a place to plug and charge your phone, laptop, or tablet.

It even shows which outlets are taken by coloring them in red, wile showing which ones are available by coloring them in green.

It's a beautiful system that would make a painful task, that much easier. This type of technology could even be extended to finding a table in a cafeteria or a several seats on a train.

Google Glass showing where the baggage claim is

The third picture shows you where your specific baggage claim is, both with an arrow in the right direction and the baggage claim number.

It also shows how many more minutes until bags start unloading. This type of technology would take a lot of the stress out of traveling through airports.

It could stream line the entire process and make the start or end of a vacation *that* much more enjoyable.

Google Glass showing how much cab fare would cost

The last picture is probably one of the more useful applications of Google Glass. It shows how much a cab ride would cost to get to your destination and how long it would take to get there.

This technology could create a healthy amount of competition between cab companies, all of them showing you their lowest prices. With you just being able to pick and choose what you want.

JetBlue is just one example of how augmented reality can used in our everyday lives. The possibilities really are endless.

That's what makes this most the most exciting aspect of glass, it's just a glimpse at the true potential. The best is yet to come.

Another example is Google Goggles; this is (in some ways) an augmented reality app that is already available for any Android smart phone.

However it's not a true augmented reality app as it needs you to take a photo of something out in the world, such as a landmark or painting, and it will then pop up information about that thing.

While a true augmented reality app would simply show you that information over your regular view, like Google Glass does. The difference is small but important.

Google Goggles paves the way for some amazing future applications of augmented reality on Google Glass.

It has a ton of cool features that would work amazing well on Glass. Have a look at some of them coming up.

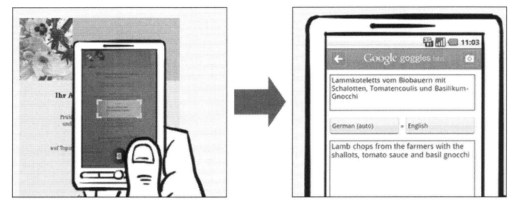
Google Goggles translating a menu

Google Goggles cantranslate a menu in a foreign country. This would make eating out at restaurants that much easier while on vacation.

This would translate over to Google Glass so easily, you would simply have to look at the menu and it would show the translations of anything you want.

In face this translating technology would work on just about any sign in a forging country.

You could use it to read information boards, or scan the headlines of the local paper. The possibilities go on and on.

Google Goggles showing information on a famous landmark

Google Goggles will let you take a picture of any famous tourist attraction or famous land mark and it will tell you all you would ever need to know about it.

This type of feature would make Google Glass yet another amazing addition to any vacation.

No longer would you need to carry around a multitude of different tourist handbooks and guiding trying to figure it all out.

You could even discover new things about landmarks and sights you didn't even know existed, just by looking at them no less.

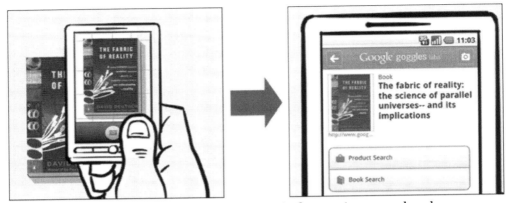
Google Goggles giving more information on a book

Google Goggles will let you take a picture of any book and tell you everything you would want to know about it.

Everything from the date it was published to the edition number, it will even show you more books by that author or similar books in that genre.

In fact you can even price check the book against other stores, both physical and online and see if you are getting the best deal in a bookstore.

This is yet another feature that would easily transition over to Google Glass. It's something that would fit it so well.

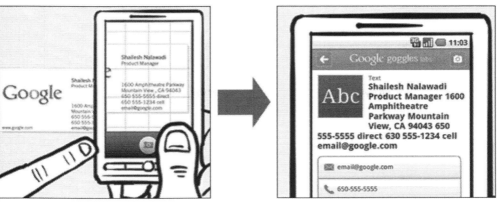
Google Goggles saving contact info

Google Goggles can take contact info right off a business card and store it on your phone; you don't even need to open the keyboard.

This would save a lot of time searching for different business cards, or even remember random phone numbers or addresses.

The best part is that there's nothing to say that Google Glass couldn't just automatically remember it for you. All you would have to do is look at it once and you're set.

From then on out that person or businesses number and address is automatically in your phone just waiting to be used by you.

Google Goggles displaying extra information about art

Google Goggles can show you relevant information about any piece of art. Who created it, when it was made, how old it is, what the inspiration was, virtually any piece of information you could possibly want to know.

This doesn't just extend to famous art pieces, the technology doesn't have any limits, and it could work for just about any piece.

As long as there is some information store about it online Google can find it.

With Google Glass discovering new art and information behind famous pieces will be easier than ever before.

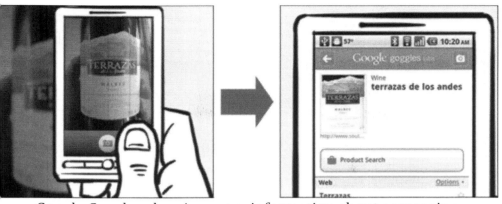
Google Goggles showing more information about some wine

Google Goggles can show you more information about the wine you could be drinking.

It could even give you reviews about that very wine and how you how other people liked it.

This would steer you clear of any terrible choices and help you save money by only buying well recommend wines.

You could even figure out which wines are similar to the very one you could be drinking.

This can extend past wines and to a while variety of other things.

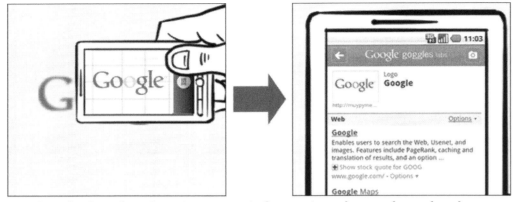
Google Goggles showing more information about a brand or logo

Google Goggles can show you everything there is to know about a brand or logo, even if it's on a pop can or on a roadside advertisement.

This will make it much easier to see if a new brand is worth your time or time.

Just how wine could have review and ratings so could brands. Entire new ways of recommending and discovering new brands would become possible.

Google Glass would make all these things so much more useful. They would be accessible at a moments notice.

Although these features do not exist on Glass right this moment, it won't be long until they become the centerpiece of what Google Glass has to offer.

Augmented reality is the future of technology; it's a way to get new and useful information to everyone quickly and easily. Augmented reality is creating the connected future.

It's the technology of the future and Google Glass will push it to new heights.

How Google Glass Handles Your Privacy and The Privacy Of Those Around You

Does Google Glass make privacy a priority?

With the front facing camera of Google Glass looking out into the world, a lot of concern over people's privacy has been brought up.

How does someone know if a person with Google Glass is filming them?

Or taking a picture?

What if Google Glass has some kind of facial recognition software and that person immediately knows who I am?

These are valid concerns, in fact they are so valid that the American, Canadian, and European governments have reached out to Google and asked for answers.

The answers so far have been strikingly clear: Google wants to protect people's privacy when it comes to Google Glass.

That means that Google Glass will not be able to use facial recognition software to figure out who's around you.

People are always looking for more privacy in the modern world

Another thing that came up is that the person using Google Glass must stare at the person if they want to take a picture or video of them.

This seems to have calmed most people's fear of what could be done with glass.

Google has tried making it as clear as possible, especially to the different governments that Google Glass has built in protection for protecting other peoples privacy.

But in the end a lot of people just won't be comfortable around Glass, as there are always around these types of settings.

Time will tell how the world reacts to Glass and what it can do.

Google Glass and The Connect World

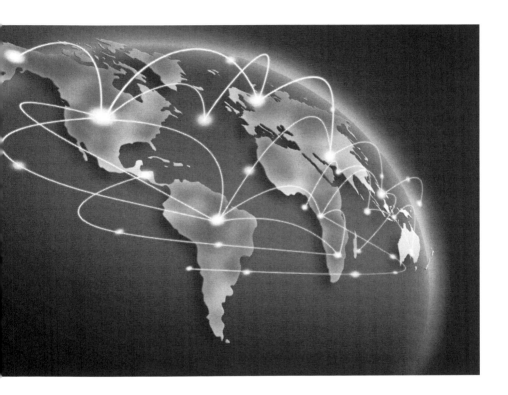

Google Glass isn't just trying to revolutionize how we interact with technology today but it plans to do so for the foreseeable future.

The connected world has already made a massive impact in virtually every corner of the modern world.

Every business and home is filled with the latest computers, smart phones, tablets; even the TVs are becoming smarter.

All of these devices are connected to the Internet; every one of them is connect to our world more and more.

Google Glass is just the next evolutionary step, another cog in the modern technology machine.

It will one day be looked back on like we now look back on the invention of the cell phone or the laptop.

Google glass is a natural extension of what technology can do and it connects our world in a way that nothing before it has been able to achieve, it separates out the technology barrier.

As smartphones become more and more common place they take up more and more of our time and energy. They begin disconnecting us from the world as much as they connect us.

Google Glass goes in the opposite direction. By removing any type of physical interface besides the display, it cleans up a lot of the clutter in our lives.

With Google Glass you can just enjoy the moment without having physical technology get in the way

It removes any thing that we have to hold, or devote our attention too. It let's us enjoy the world around us more.
That clutter was slowly disconnecting us from the people and world around us.

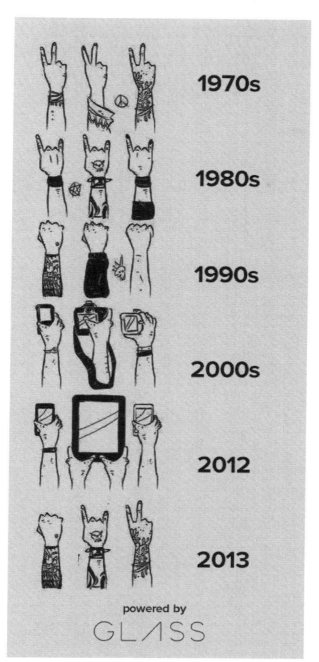

Meanwhile Glass makes the simple tasks of a smartphone or tablet faster and more efficient.

With this efficiency we can accomplish more, we can learn more, connect more, and at the same time spend less of our lives hooked on to our technology.

That's the true difference that Google Glass offers us over the previous advancements in technology and that's not even going into the whole new world that augmented reality offers us either.

Google Glass was build with sharing in mind, the amazing set of hands free camera and video apps are a testament to that.

As are the different social media apps such as Facebook and Twitter, with Glass it has never been easier to share a photo with your friends.

But Google Glass goes even further than that. Because it's so accessible, as it's always right there ready to record or snap a picture, sharing of moments will become a much more common thing.

Footage of a mountain bike race as seen from the perspective of the rider

It could become the next GoPro of the everyday world.

Sharing moments from the sideline of a kids soccer game to being in the stands for the final lap of a formula one race.

The overall impact will be much greater than what we have seen with smart phones, as so many moments are just not easy or practical to film with a smart phone.

A photo taken from the sidelines of a formula one race

The benefit that Glass has it that it's always ready and always connected.

Another feature that Glass brings to the table is being able to show anyone who you're having a video chat with exactly what you see.

You could share your kids first steps, or words. You could even teach the other person how to cook or play golf just by showing them what you see.

Sharing these moments with another person is simple and easy.

Google Glass can easily message it to someone, email it, or even unload it to a variety of social media websites.

All of it happens seamlessly and without even pressing a single button.

Google Glass looking stylish

The possibilities are practically limitless and the best part is it's always right there; completely hands free and ready to go.

That's the true power of sharing with Google Glass.

What Goes Into Glass?

You might be wondering just how Google Glass stacks up to your smart phone or tablet. Well here is everything you'll need to know.

Google Glass is equipped with a 5MP camera, which is considered about average as far as smart phones are concerned.

The camera also comes equipped with software to take good low light photos.

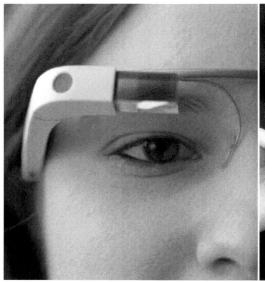 

Glass' camera can be seen as the gray dot to the right of the lense

Glass runs on a 1.01GHz duel core processor. This is roughly the same as any top of the line smartphone.

It also has 1GB of Ram of 12GB of storage.

The display is truly one of a kind. The average smart phone has a pixel density of around 326 per inch (for the iPhone 5). While Glass has a density of around 3337!

Google Glass' display is truly one of a kind

More than 10 times what the iPhone has. That's of course because of how small Glass' display is and how close it really is to your eye.

It comes equipped with WiFi and Bluetooth 4.0. Unfortunately it does not come with GPS or any sort of cell connection, so you will need to pair it with your smart phone.

Google Glass is Only Just Beginning

The story of Glass is only now being written, we are about to enter a new age of technology where everything is connected and the barriers of technology begin to be lifted away.

The feature set of Google Glass will only get stronger as time goes on and as new technologies are developed.

The advances we see in the up coming years might just blow everything you've read in this book out of the water.

Who knows, maybe augmented reality on Glass is bigger than we could have ever dreamed; only time will tell.

This marks the beginning of Glass and the Future.

Made in the USA
San Bernardino, CA
21 July 2014